T0400720

CUTTING-EDGE TECHNOLOGY

ALL ABOUT NANOTECHNOLOGY

by Racquel Foran

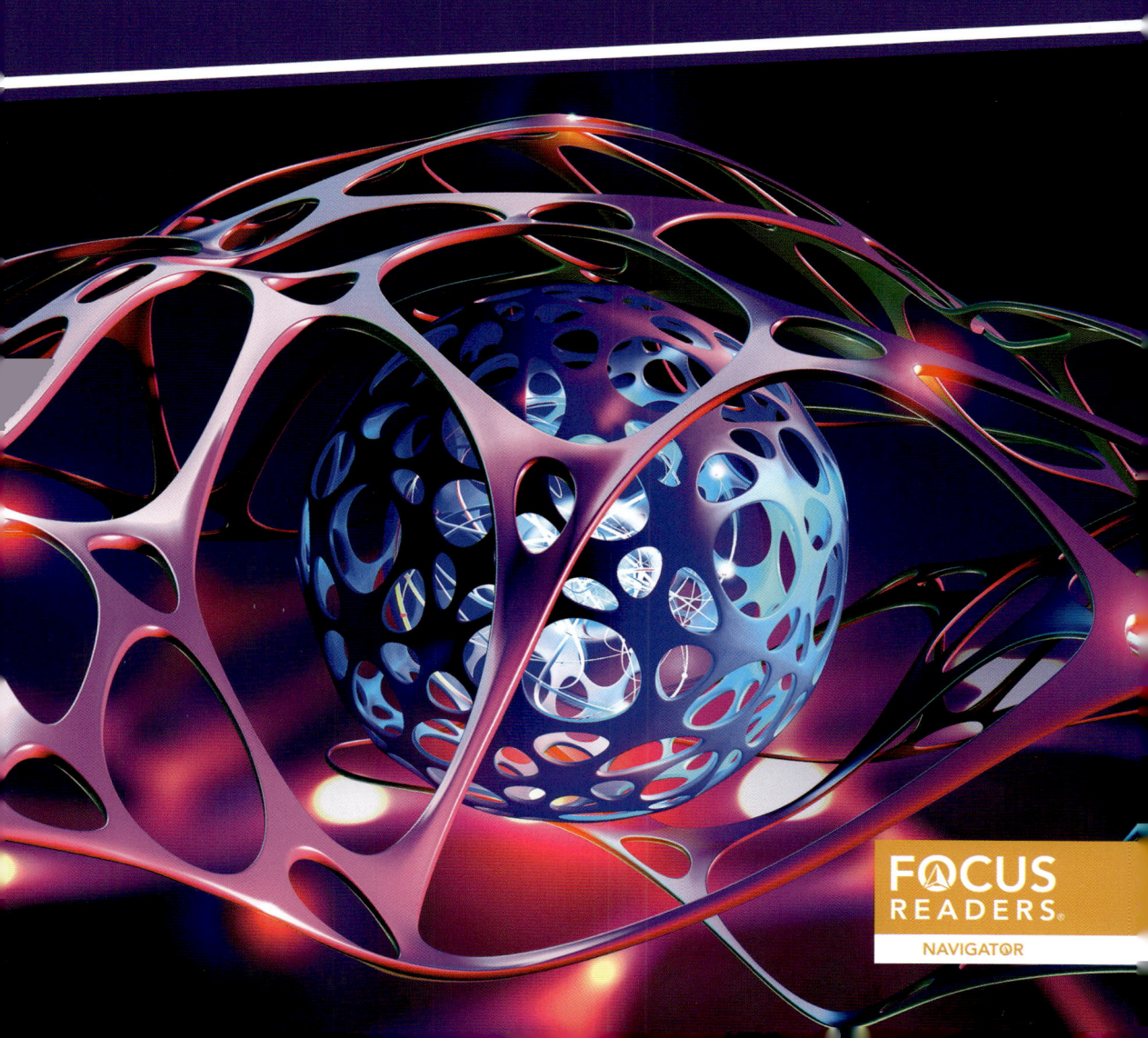

FOCUS
READERS.
NAVIGATOR

WWW.FOCUSREADERS.COM

Focus Readers is distributed by North Star Editions:
sales@northstareditions.com | 888-417-0195

Produced for Focus Readers by Red Line Editorial.

Content Consultant: Muhammad M. Rahman, PhD, Assistant Research Professor, Department of Materials Science and NanoEngineering, Rice University

Photographs ©: Shutterstock Images, cover, 1, 6, 8–9, 10, 17, 23, 29; Martyn F. Chillmaid/ Science Source, 4–5; National Institute of Standards and Technology/Science Source, 12–13; James King-Holmes/Science Source, 14, 25; McComas/MCT/Newscom, 19; Brookhaven National Laboratory/Science Source, 20–21; NASA/Pat Rawling/Science Source, 26–27

Library of Congress Cataloging-in-Publication Data
Names: Foran, Racquel, 1966- author.
Title: All about nanotechnology / by Racquel Foran.
Description: Lake Elmo, MN : Focus Readers, [2023] | Series: Cutting-edge technology | Includes bibliographical references and index. | Audience: Grades 4-6
Identifiers: LCCN 2022031724 (print) | LCCN 2022031725 (ebook) | ISBN 9781637394731 (hardcover) | ISBN 9781637395103 (paperback) | ISBN 9781637395806 (ebook pdf) | ISBN 9781637395479 (hosted ebook)
Subjects: LCSH: Nanotechnology--Juvenile literature.
Classification: LCC T174.7 .F66 2023 (print) | LCC T174.7 (ebook) | DDC 620/.5--dc23/eng/20220822
LC record available at https://lccn.loc.gov/2022031724
LC ebook record available at https://lccn.loc.gov/2022031725

Printed in the United States of America
Mankato, MN
012023

ABOUT THE AUTHOR

Racquel Foran is a freelance writer from Coquitlam, British Columbia, Canada. She has authored several nonfiction titles for school-age readers on diverse subjects, including organ transplants, autism, and North Korea. When she is not writing, she enjoys tending to her Little Free Library, painting, and hiking by the river with her dogs.

TABLE OF CONTENTS

CHAPTER 1

TINY TECH

Y ou buy a new pair of glasses from a store. You put them on and walk outside. At first, you blink and squint because the sun is shining brightly. But soon you can see clearly. That's because the lenses are photochromic. They change color in response to light.

By changing color in bright light, photochromic lenses protect the wearer's eyes.

 Nanofilms can keep dirt and water from sticking to lenses. The films also change how lenses reflect light.

Inside the store, the lenses were clear. However, they turned dark when the bright sunlight hit them. Changing color helps the lenses block **ultraviolet** light. A clear nanofilm on the lenses makes this happen. This super-thin film also helps the lenses avoid scratches. It stops them from fogging up, too. It even keeps the lenses clean.

Nanofilm is a kind of nanotechnology. Nanotechnology uses science to make and control extremely tiny things. These things are too small for people to see with their eyes. But the tiny things have many uses. Nanotechnology can improve medicine, electronics, and much more.

STOPPING FOG

Eyeglass lenses can fog up when temperatures change. Tiny drops of water collect on the lenses. They make it hard for people to see. Scientists have made a nanofilm that can help. The film has tiny bits of gold in it. These particles absorb light from the sun. The light warms the lens, which keeps droplets from forming.

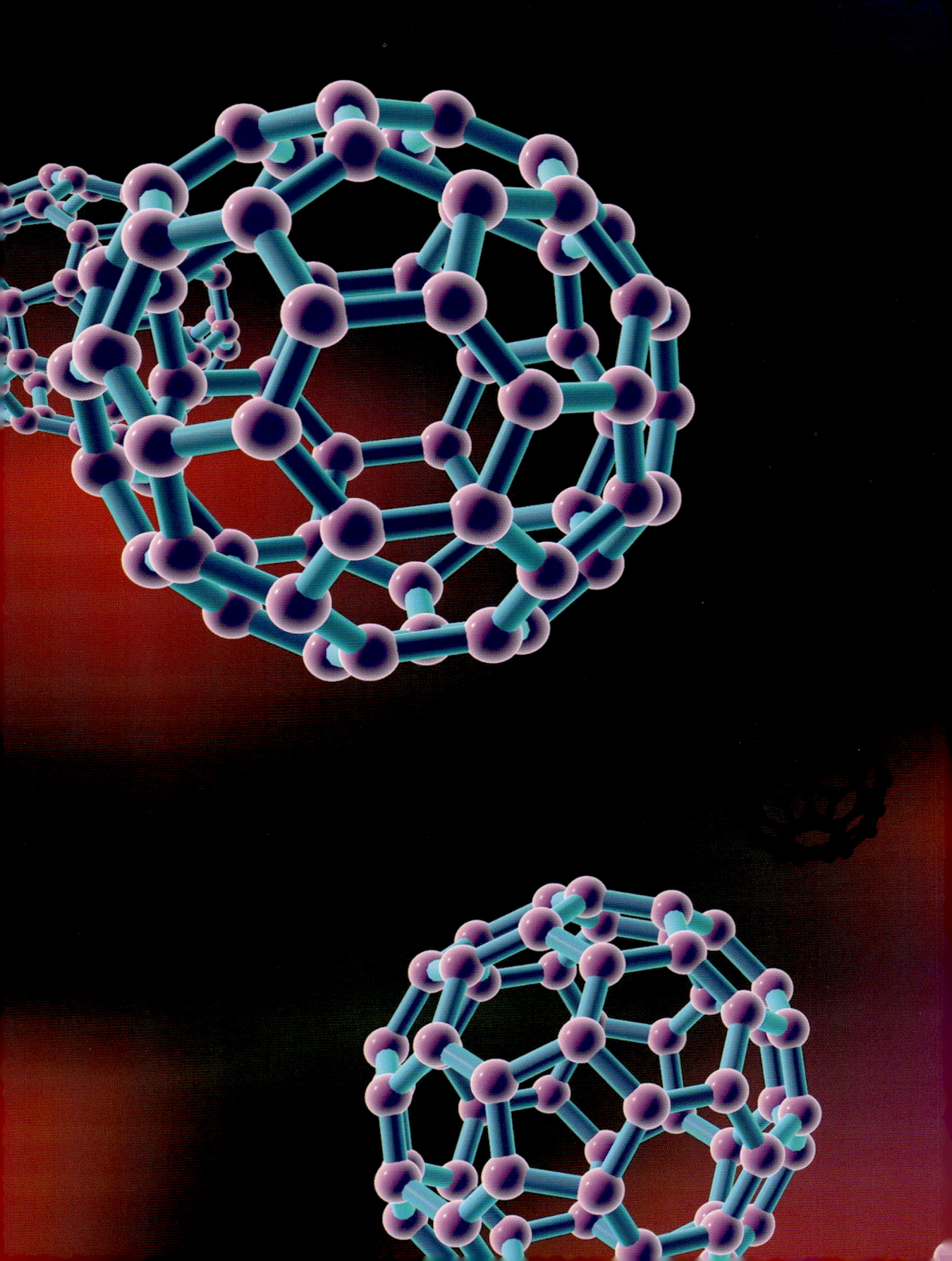

CHAPTER 2

NANOPARTICLES

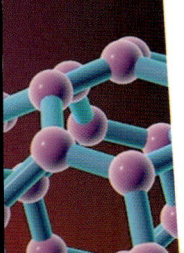

Nanotechnology uses tiny bits of matter. These bits are measured on the nanoscale. This scale uses units called nanometers. One nanometer is equal to one-billionth of a meter. That's incredibly small. A piece of paper is approximately 100,000 nanometers thick.

Some nanoparticles are one million times smaller than an ant.

Many nanoparticles are made from carbon. This dark-colored element is good at forming strong bonds.

Scientists use the nanoscale to measure **atoms** and **molecules**. When atoms join together, they form something new. Nanoparticles are one example. These particles are still tiny. They are between 1 and 100 nanometers. But they can contain hundreds of atoms.

There are many types of nanoparticles. Scientists study how these particles look and behave. Some are flat. Others look like boxes, balls, or tubes. Scientists also work with nanomaterials. These materials form when nanoparticles join together. Some are small enough to measure in nanometers. Others are larger. But they tend to be very thin.

NANOMATERIALS

Some nanomaterials are natural. Smoke from fire is one example. However, many nanomaterials are created by people. Scientists design them to have certain traits. For instance, people can create strong fibers or hollow tubes. These materials improve many types of technology. Some make protective coatings. Others give doses of medicine.

CHAPTER 3

NANOTECHNOLOGY HISTORY

For a long time, nanomaterials were too small for people to see. That changed in the 1980s. Two new microscopes were invented. One was the scanning tunneling microscope (STM). It was created in 1981. It uses electricity to create very close-up images. Scientists could see individual atoms. But they could see only

The STM moves a metal tip over a surface. By sending out electricity, it can find a single atom.

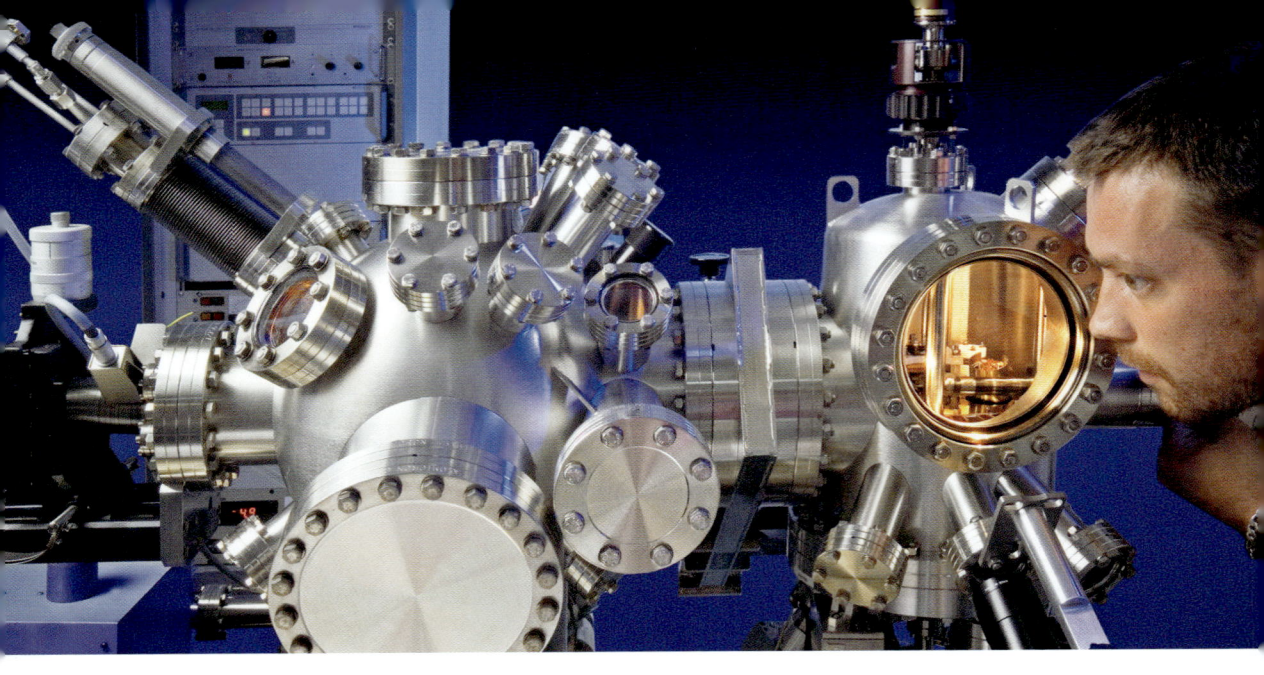

An AFM moves across a surface and creates a very close-up map.

some types. The atoms had to **conduct** electricity.

The atomic force microscope (AFM) was created in 1986. It didn't rely on electricity. So, it worked for more types of materials.

With these new tools, scientists found new particles. Buckyballs were discovered in 1985. These nanoparticles are made from carbon. They are shaped like balls.

In 1991, a scientist found tube-shaped nanoparticles. Like buckyballs, they were hollow. And they were made of carbon. They became known as nanotubes.

The new microscopes also let scientists move one atom at a time. This was an exciting achievement. It meant that scientists could make new types of molecules. They could add or take away atoms. Or they could create new shapes.

Changing a molecule's size or shape can affect how it behaves. Scientists saw this with nanoparticles. Tiny bits of matter often act different than larger particles, even when they're the same material. For example, splitting gold into tiny bits changes its color. The bits become red and purple. That happens because the tiny particles reflect light differently. Smaller particles also have a high surface-to-volume ratio. Most

BETTER SUNSCREEN

Many sunscreens have minerals in them. These minerals help reflect the sun's light. To do this work, the minerals must stay on the skin's surface. They make skin look white. But some sunscreens break the minerals into nanoparticles. These particles still reflect light. They also absorb and scatter light. As a result, the sunscreen is invisible. The particles also help the sunscreen's protection last longer.

Because they are so thin, sheets of graphene are very flexible and nearly clear.

of their atoms are on the surface. That changes how they interact with things around them.

In 2003, researchers created a sheet of carbon that was one atom thick. They called it graphene. It could be used to make tiny machines.

GRAPHENE

Graphene is the world's thinnest material. It's also the strongest. It has a high surface-to-volume ratio. And it conducts both heat and electricity. As a result, graphene can be used in a variety of ways. Many uses involve wearable devices. These are small machines people wear. A smartwatch is one example. But devices made with graphene tend to be much smaller.

Scientists are also developing temporary tattoos made from graphene film. Graphene film is very flexible. It can stretch to fit any shape. And it can wrinkle with skin. So, it's good at sticking to a person's body.

Nano-size sensors can be applied to these tattoos. The sensors can collect data about the wearer's health. For instance, they can measure the person's heart rate. Muscle activity and brain

STICK-ON SENSORS

A thin film made of graphene sticks to the wearer's skin.

Tiny sensors pick up changes in movement, heat, or electricity.

A tiny radio transmitter sends these signals to other devices.

Thin wires make a circuit that can bend without breaking.

waves can be tracked, too. The tattoos are easy and cheap to make. They could become a key way to track medical information.

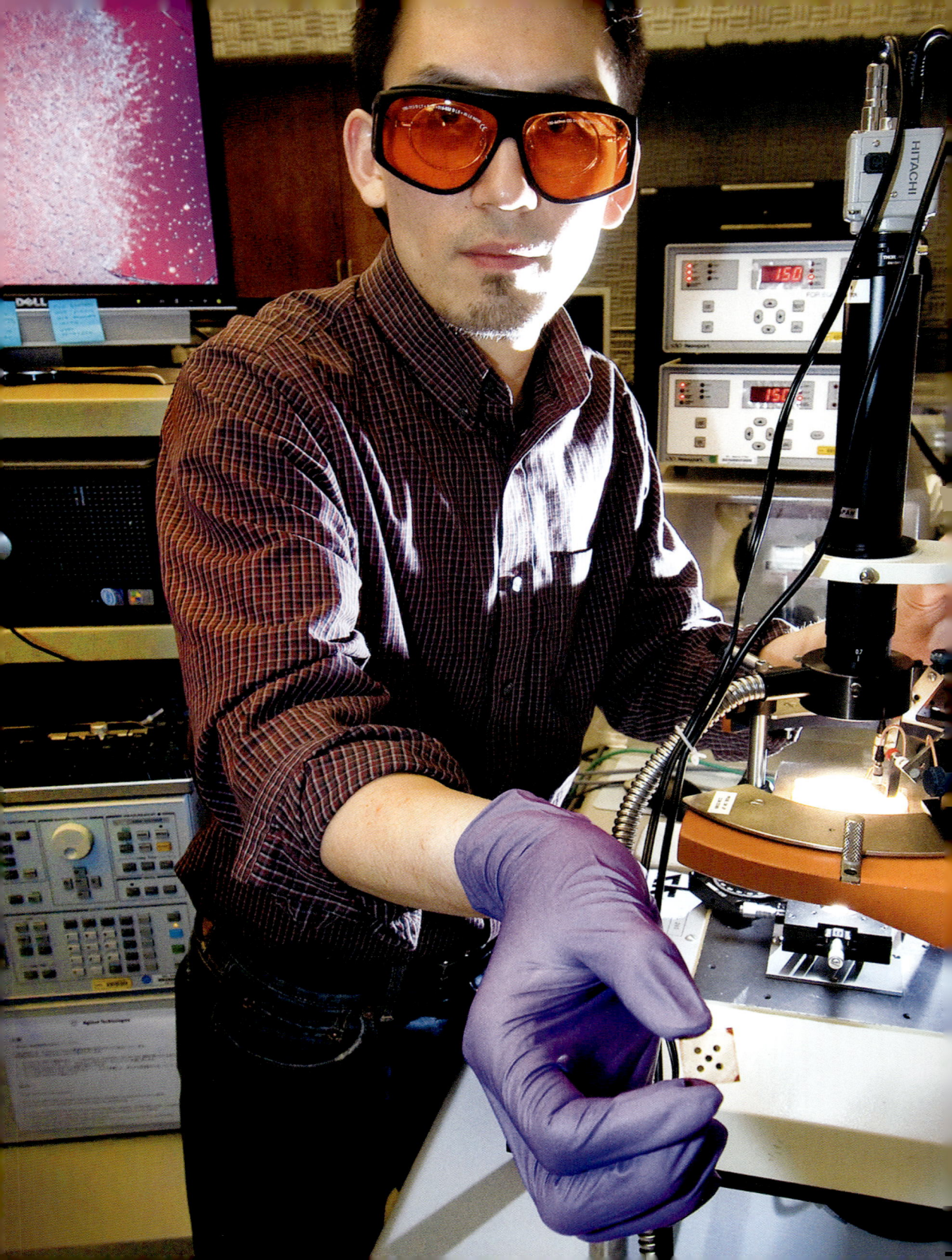

CHAPTER 4

USES OF NANOTECHNOLOGY

Nanotechnology allows scientists to improve many materials. For example, plastic can be made stronger and lighter. That can reduce the cost of space travel. Lighter rockets use less fuel.

Nanomaterials have improved sports equipment, too. They make tennis rackets stronger. And they help balls last longer.

Scientists at the Center for Functional Nanomaterials test new uses for nanotechnology.

Nanoparticles are useful in clothing as well. For instance, some nanoparticles make fabric stronger. Others help fabric stretch. Nanoparticles can even help clothing smell better. Some companies put silver nanoparticles in their fabric.

NANOCOATINGS

When ships move through water, plants and algae stick to their sides. That makes the ships move more slowly. So, they must use more fuel. Airplanes face problems, too. During cold weather, ice can form on their wings. Planes cannot fly when that happens. Nanotechnology can help in both of these cases. Scientists can coat vehicles with materials that include nanoparticles. These coatings keep things from sticking to the vehicles.

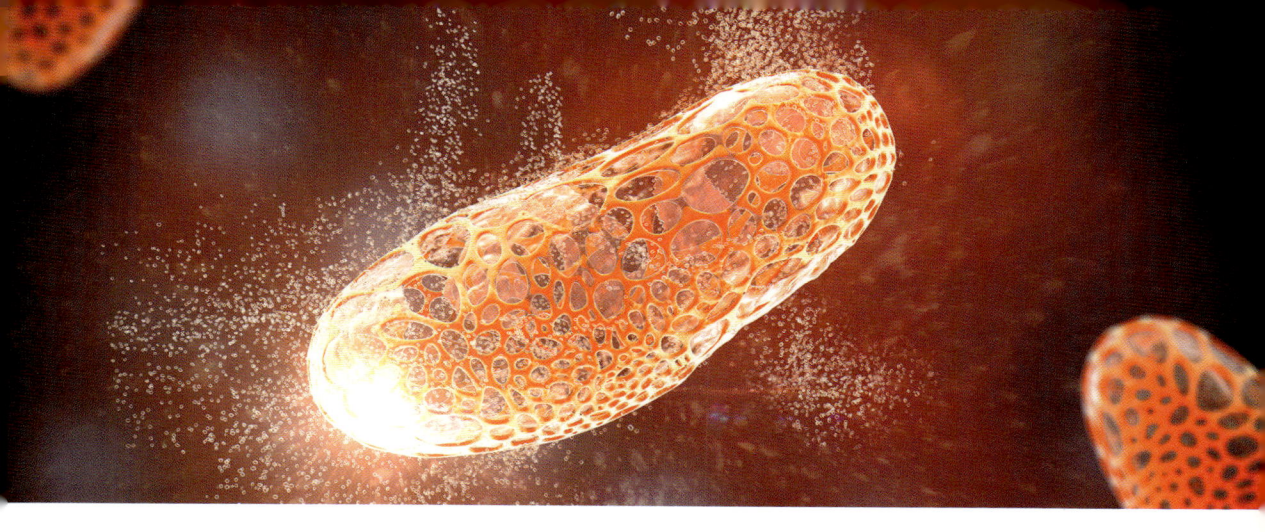

Nanoparticles could help find and fight the bacteria that cause some diseases.

These particles have **ions**. The ions kill the **bacteria** that make clothes stink.

Doctors also use nanotechnology. It helps them treat cancer. Scientists add nanoparticles to medicines. The particles are attracted to cancer cells. They target those cells directly. Healthy cells are not affected. So, the person won't feel as sick.

Nanoparticles play an important role in the food industry. Some particles make food packaging stronger. Others help slow or prevent the growth of bacteria. By doing this, they help food last longer. Nanoparticles could also make food healthier. Adding them to food could help people get more nutrients.

Other types of nanotechnology are used to make tiny devices. For example, many devices use circuit boards. These small, flat boards connect electronic signals. Computers, cars, traffic lights, and fridges all use them. Nanotechnology allows more parts to fit on one board. The boards take up less space. And they can be more powerful. Electronic devices can be smaller as a result.

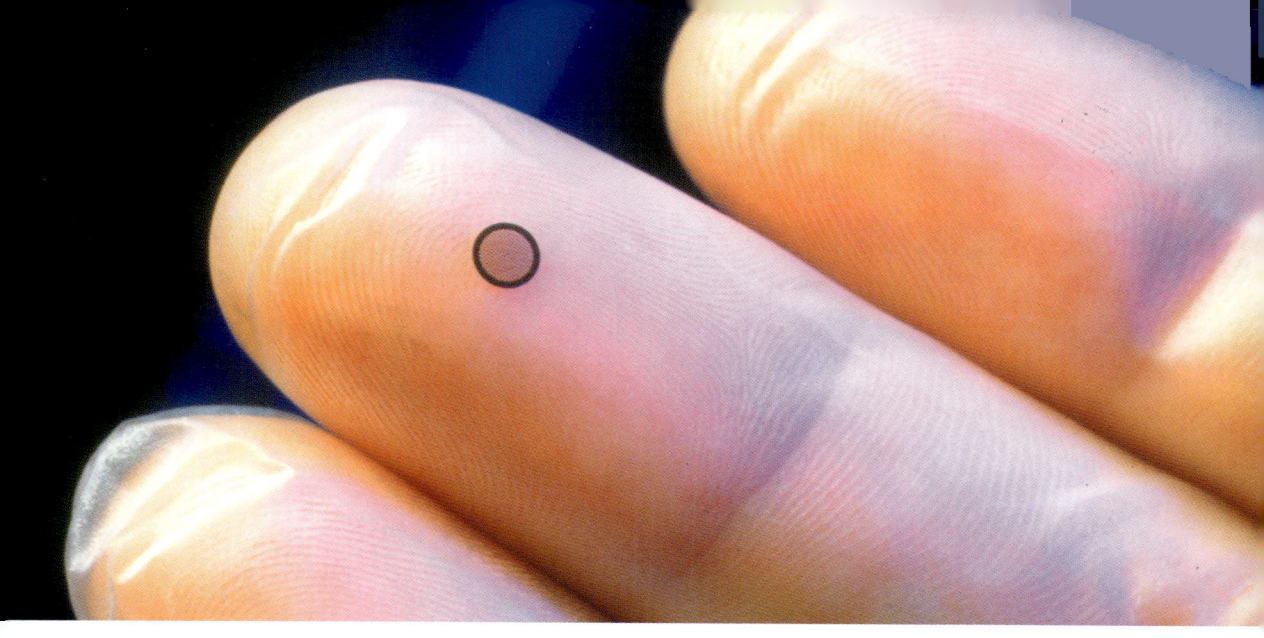

This dot is 50 nanometers thick. Using nanotechnology, all the information in the *Encyclopaedia Britannica* could be written on it.

Nanotechnology can even help make water cleaner. Water can become polluted with dangerous chemicals. They make it unsafe to drink. However, nanoparticles can create a reaction with these chemicals. The reaction makes the chemicals harmless. The water becomes safe again.

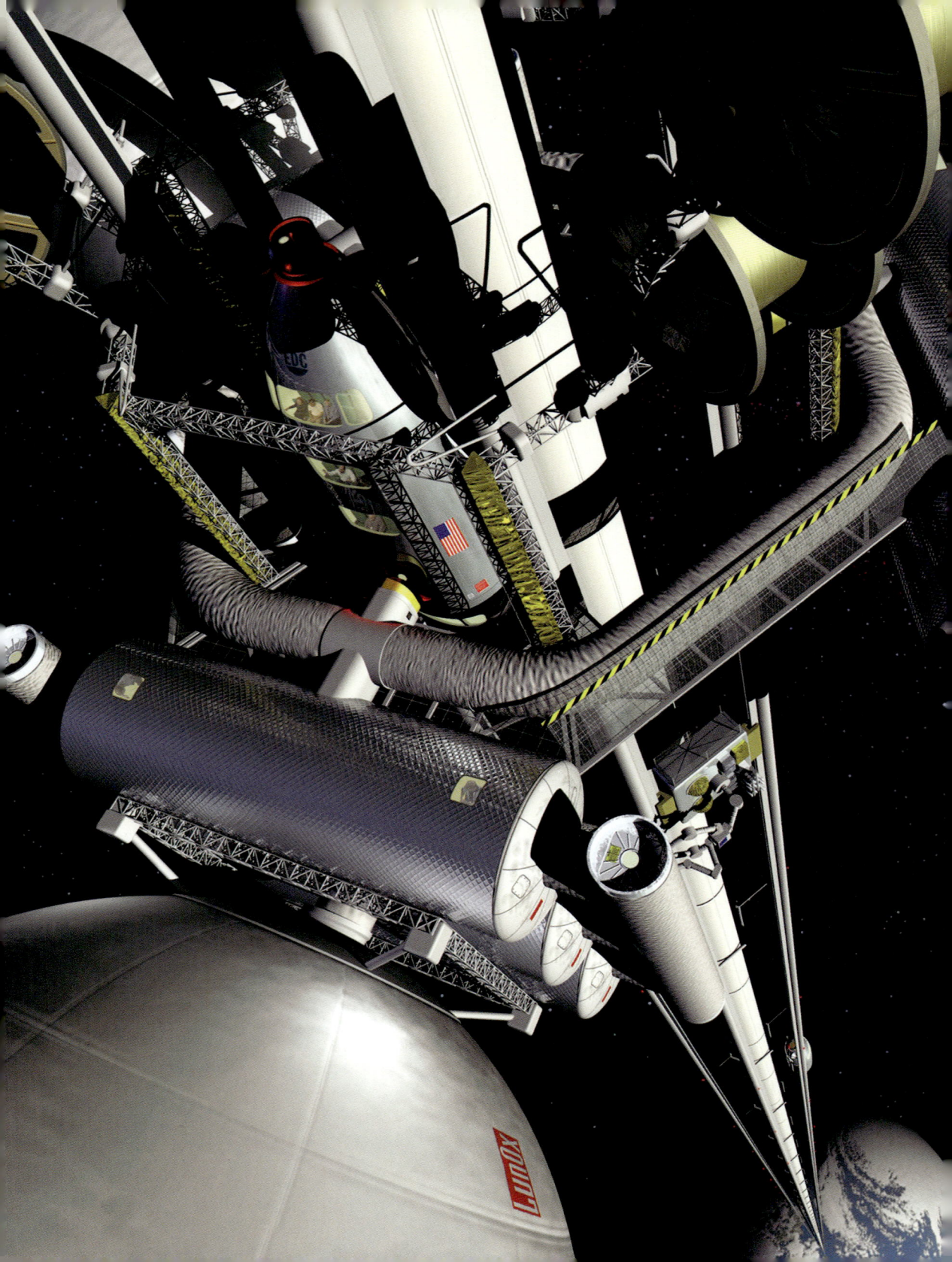

CHAPTER 5

FUTURE PLANS

Scientists see many possible uses for nanotechnology. Some scientists want to build an elevator to space. They plan to use carbon nanotubes. First, they would assemble the nanotubes into fibers. Then they would weave fibers to make cables. The project would need 144,000 miles (232,000 km) of nanotubes.

A space elevator would use a long cable to connect Earth with a space station.

Nano sensors could help with safety as well. People could print sensors on large plastic sheets. They would apply these sheets to bridges or buildings. The sensors would detect cracks or other dangers. If they sensed a problem, they would send a warning to a computer.

Scientists have also begun developing nanobots. These tiny machines could be injected into people's bodies. They could give medical care from the inside. Some may repair damaged **tissues**. Others may clear clogged **arteries**. That could help prevent strokes or heart attacks.

Nanotechnology can help fight **climate change**, too. Many electric car batteries

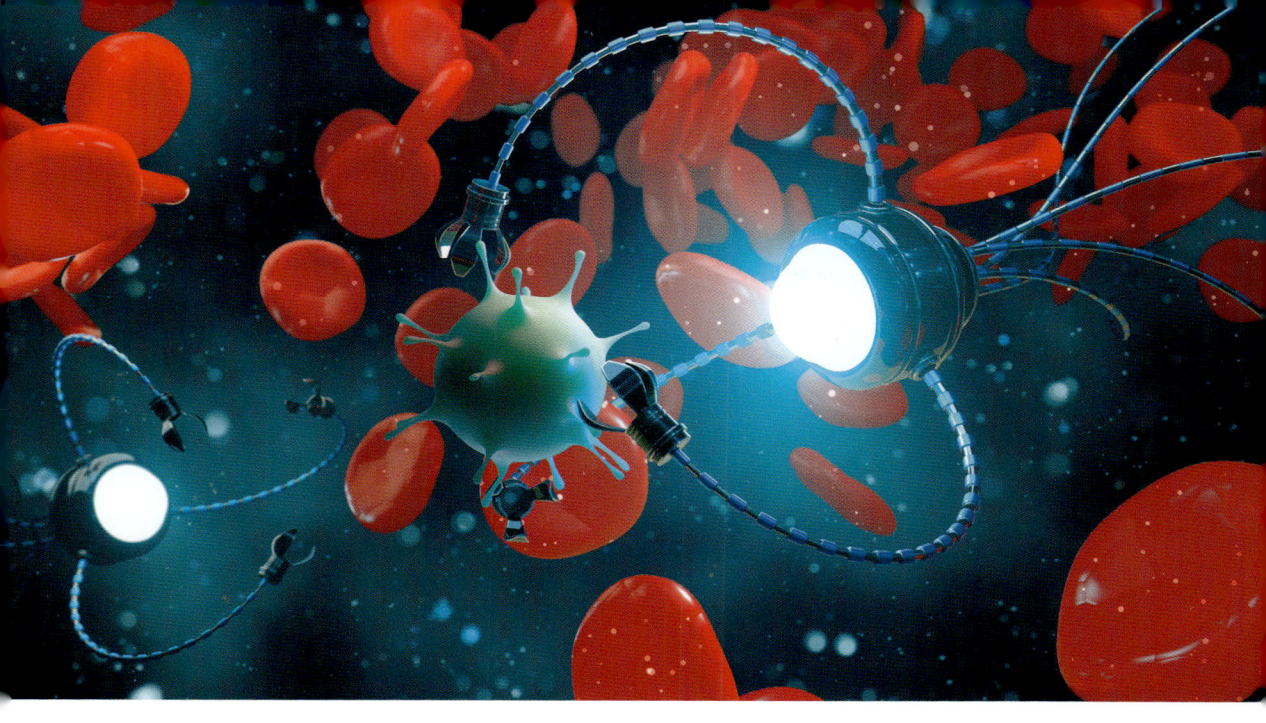

Nanobots are between 50 and 100 nanometers wide. Some are tiny machines. Others are made of living cells.

use lithium. Graphite powder often joins this material together. New batteries use carbon nanotubes instead. The nanotubes can join more lithium. They make the batteries last longer and store more energy. Scientists are even studying how nanotechnology could create energy.

FOCUS ON
NANOTECHNOLOGY

Write your answers on a separate piece of paper.

1. Write a paragraph explaining the main ideas of Chapter 3.

2. Which use of nanotechnology do you think is most exciting? Why?

3. How small is one nanometer?
   - **A.** one-billionth of a centimeter
   - **B.** one-billionth of a meter
   - **C.** 100,000 centimeters

4. How would using nanotechnology to improve batteries help fight climate change?
   - **A.** The new batteries wouldn't use or store energy.
   - **B.** The new batteries would help people use and store cleaner energy.
   - **C.** The new batteries would force people to stop using energy.

Answer key on page 32.

GLOSSARY

arteries
Tubes through which blood flows from the heart to the rest of the body.

atoms
The smallest building blocks of matter. They make up everything in the physical world.

bacteria
Microscopic, single-celled living things.

climate change
A human-caused global crisis involving long-term changes in Earth's temperature and weather patterns.

conduct
To allow heat or electricity to move through something.

ions
Atoms or groups of atoms that carry an electric charge.

molecules
Groups of atoms that are joined together.

tissues
Groups of cells in the body that have certain functions.

ultraviolet
A type of invisible light that can cause sunburns and cancer.

TO LEARN MORE

BOOKS

Kulz, George Anthony. *Nanotechnology*. Chicago: Norwood House Press, 2018.

Rathburn, Betsy. *Nanotechnology*. Minneapolis: Bellwether Media, 2021.

Schwarz, Venessa Bellido. *Medical Technology Inspired by Nature*. Lake Elmo, MN: Focus Readers, 2019.

NOTE TO EDUCATORS

Visit **www.focusreaders.com** to find lesson plans, activities, links, and other resources related to this title.

INDEX

Answer Key: **1.** Answers will vary; **2.** Answers will vary; **3.** B; **4.** B